A GUIDE
MOTOR CYCLE DESIGN

A COLLECTION OF
VINTAGE ARTICLES ON
MOTOR CYCLE CONSTRUCTION

By

VARIOUS AUTHORS

Read & Co.

Copyright © 2022 Read & Co. Books

This edition is published by Read & Co. Books,
an imprint of Read & Co.

This book is copyright and may not be reproduced or copied in any
way without the express permission of the publisher in writing.

British Library Cataloguing-in-Publication Data
A catalogue record for this book is available
from the British Library.

Read & Co. is part of Read Books Ltd.
For more information visit
www.readandcobooks.co.uk

CONTENTS

3

These articles have been extracted
and compiled from various editions of
The Model Engineer and Electrician.

A GUIDE TO
MOTOR CYCLE DESIGN

WHY THE CRANK CASE GETS HOT

The motor cyclist is often puzzled to know why the crankcase of his engine gets so hot, a condition which is almost certainly accompanied by a falling off in power and speed. I have at various times received enquiries from readers complaining of these symptoms and in one case, a novice rider told me that he contemplated either removing the drain plug or else drilling a hole in the top of the crankcase to "allow of the heated air escaping." Fortunately, I was in time to prevent him from doing considerable injury to his engine by adopting either of these plans.

A hot crankcase—that is, one which gets unusually warm— indicates that the piston rings are in a faulty condition, and that, owing to their lack of "fit" in the cylinder, gas is leaking past them into the crank chamber. This results not only in conveying heat to the latter, but also in loss of power, and the remedy is to fit new rings. On inspection they will most likely be found to bear marks of discolouration, which is a sure sign that leakage is taking place.

Sometimes, if the rings are removed from the piston and the grooves cleaned out, the inside of the rings themselves being also scraped clear of burnt deposits, some further use may be made of them, but as a general rule it is better to fit new ones If the cylinder has had a very great amount of wear it may be necessary to have it re-bored, so as to restore its true cylindrical interior shape, and then a new piston altogether will be required.

A MAGNETO DIFFICULTY

A Weymouth reader, who possesses a motor bicycle fitted with U.H. magneto, writes, asking my advice in the following circumstances: "I experienced during a recent ride," he says, "a most peculiar action on the part of the magneto. So long as the timing lever was in the fully advanced position or any position between that and, say, two-thirds retarded, all was well. Movement of the lever further than this, however, in the direction of retard immediately brought about misfiring, whilst fully retarding it was to stop the engine firing altogether.

"It is a low-powered machine, and I find it necessary to fully retard the ignition when climbing a stiff hill, but as the effect of this was to stop the engine, I was obliged to desist, and knocking and labouring of the engine resulted.

"On one long down grade I used the timing lever as a switch, cutting off the current to the sparking plug by fully retarding the lever."

My view of the matter is that the contact breaker points required adjusting. In the U.H. magneto this can only be done with the parts *in situ* and a special spanner is provided for the purpose. This appliance was illustrated some time back in these pages.

Immediately at the rear of the contact disc is a rotatable washer or plate, moved by means of the special spanner referred to. If this plate be turned to the right the effect is to reduce the amount of "break" and to retard the moment of sparking, whilst if, on the other hand, it be turned to the left, the "break" is increased and the spark advanced. Presumably, owing to want of adjustment, the "break" was taking place too late in the stroke and the point of maximum retard was reached with the lever still some short distance away from the point of retard in its travel.

ELEVEN YEARS OF
MOTOR CYCLE DEVELOPMENT

The past ten or eleven years have wrought a wondrous difference in motor cycle design and construction, and indeed there is little in common, unless it be the main principle of working, between the machine of to-day and that of a decade ago. Yet those of us who, as motor cyclists, date Our experiences back to the period of 1902 or thereabouts, can recall many a pleasurable ride on machines which in these later days would be looked upon as curiosities and as belonging to a time when, as was generally supposed, only the hardiest ventured forth on a motor cycle, and then without any certain prospect of returning by the same method.

THE FIRST F.N. MOTOR BICYCLE

THE F.N. 4-CYLINDER 7 H.P. MOTOR BICYCLE FOR 1914.
THREE SPEEDS, FREE ENGINE
CLUTCH, AND KICK STARTER.

The development of the motor cycle has been extraordinarily rapid, and it is only since machines were brought to something akin to their present state of perfection that the motor cycling movement has assumed the huge popularity which it now enjoys. I often wonder, when present at a motor cycling meet, or at the Olympia show, or when watching the almost constant procession of riders along a main road on a fine Saturday or Sunday afternoon, how many of these later-day enthusiasts would have taken up motor cycling under the old conditions when, as a matter of course, machines were less reliable and conditions generrally less favourable. I think it may be safely assumed that the percentage would have been a small one. Certainly there would have been no likelihood, as there undoubtedly is at present, of the motor cycle becoming almost as common as the pedal cycle, for it would seem that one day this is bound to be the case.

A CONTRAST IN MOTOR CYCLES

Perhaps no better means of contrasting the motor bicycle of ten or eleven years ago and that of the present day could be selected than the one which I am enabled, by the courtesy of Mr. H. G. Bell, Sales Manager of the Motor Cycle Department of F.N. (England), Ltd., to illustrate herewith.

The belt-driven machine was built in 1902, and found a ready sale, both in this country and abroad. It had a single cylinder, F.N. engine, 52 mm. bore by 57 mm. stroke, with large outside flywheel and a flat belt pulley between the flywheel and the crankcase. The belt was 1 in. wide in the form of a flat leather strap, and the belt rim on the back wheel was made of hard wood. This provided a quite satisfactory means of transmission and one which gave little or no trouble. The machine was fitted with a spray carburettor, designed and made, as was the complete bicycle, at the F.N. factory in Belgium, and Mr. Bell tells me that it was the first motor bicycle to be fitted with this type. In the early days of the motor cycle some thousands of these carburettors were sold to manufacturers of motor cycles, and, in fact, 75 per cent. of the machines back in 1902–1905 were fitted with them. The ignition was by accumulator and coil.

This was the F.N. motor bicycle of 1902. That of the present time, in one of its latest forms, is shown in comparison with it on the preceding page. Four-cylinders, developing a rated horse-power of 7–9 and transmitting the power by means of a cardan shaft with bevel gears, encased and running in lubricant, three-speed gear box, clutch in the flywheel, kick starting device, magneto ignition, mechanical lubrication, automatic carburettor, well-sprung forks, and comfortable saddle, combine as salient features of the design to produce the maximum of comfort, reliability and power, and to justify at the same time the makers' claim to have produced a "car on two wheels."

BELT SIDE OF THE FIRST F.N. MOTOR BICYCLE

3 1/2-H.P. TWINS

A correspondent asks me to voice an opinion in these notes as to the relative merits of the 3 1/2 single-cylinder motor cycle, and a twin-cylinder machine of the same horse-power. Personally, if I were considering which to purchase I should incline strongly towards the twin, for there is no comparison in my view between the steady purr of the multi-cylinder engine with its quick acceleration, and the thump, thump of a single, with its liability to knock at low speeds; the advantage being with the former type in both respects. On the other hand, the single scores somewhat heavily in other directions. It is more simple in construction, there is less to keep in order, and its maintenance entails less expense to its owner. Moreover, it is easily kept in tune as compared with the twin-cylinder, and if, in the latter, one cylinder falls off in tune, a large proportion of the power is gone, whilst at the same time, the sound cylinder is caused to work at a disadvantage owing to the drag exerted by the faulty condition of its companion.

The smoothness in running of a twin-cylinder engine is, however, a very strong point in its favour, and to those who are really earnest in regard to motor cycling, and ready to look well after their mounts, there should be no difficulty but rather a pleasure about keeping the engine in tune, and both cylinders doing an equal share of the work. Personally, as I said before, I should choose the twin, whilst at the same time recognising the virtues of the single; but I should be extremely careful to select a machine in the design of which accessibility as regards the engine and avoidance of unnecessarily small detailed parts had been observed. There are many which possess these characteristics, but others in which they are lacking. The latter I should steer well clear of.

A TRYING EXPERIENCE ON THE ROAD

A trying experience fell to my lot during a recent ride on a bitterly cold night, but, as I have gained some further knowledge respecting the engine fitted to my machine as a consequence, I am not disposed to grumble. It happened that I had to make a journey of some fifty odd miles after nightfall, and on the afternoon of the same day I discovered, quite accidentally, that one of the studs which serve, in the Connaught two-stroke engine, to hold the manifold chamber in place, and which studs are fixed in the cylinder, was broken. Not having time to attend to the matter myself, I handed the machine over to a mechanic, with instructions to remove the stump end of the stud by drilling it out, and to fit a new one in place, the job being quite a simple one which could be done with the cylinder *in situ*.

On calling for the machine, immediately before being due to start on the outward journey, I found to my pleasure, not to say astonishment, that it was ready and waiting for me, and as the engine started up at once and ran as well as ever, I was soon careering along northwards at a good pace. After covering some ten miles or so, and when well out in the open country, I noticed that the noise of the exhaust had become much louder, and at the same time there was a marked falling off in power. At first I thought that one of the exhaust pipes had worked loose— although that would not affect the power adversely—but, as matters got worse and it became necessary to change down on a hill that I usually take "on top," I fell to the conclusion that the manifold chamber must have slackened, or that the jointing between it and the cylinder had in some way become defective.

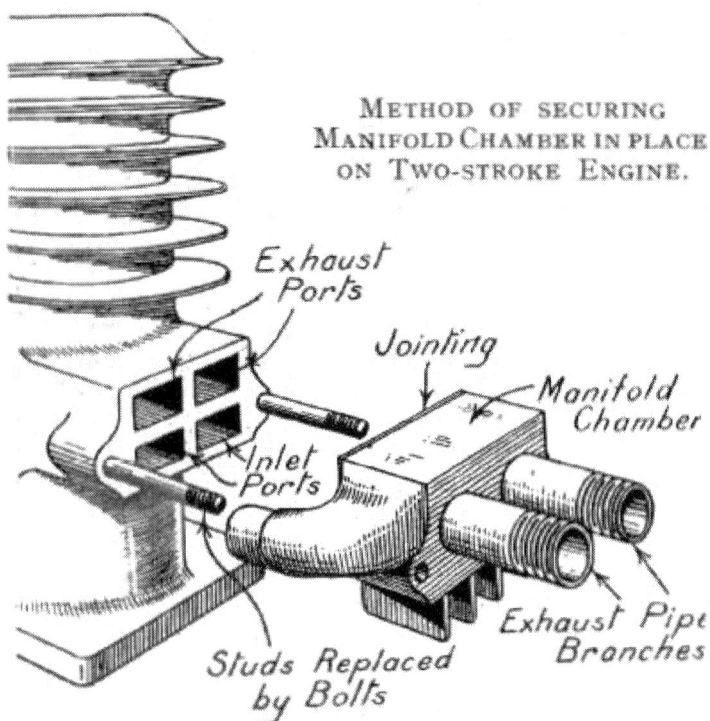

METHOD OF SECURING MANIFOLD
CHAMBER IN PLACE ON TWO-STROKE ENGINE

THE CAUSE OF THE TROUBLE

On dismounting I speedily found the last surmise to be the correct one, and whilst cogitating as to whether I should push on as best I might until reaching a point at which light and shelter would be available, or face the alternative of remedying matters on the spot, I noticed that instead of replacing the broken stud by another, the faithful workman had substituted a bolt, which was screwed into the cylinder from outside (see sketch); not only this, but he had wrought the same change on the opposite side where, so far as I knew, there was no need to meddle with things at all. However, there was one advantage to be derived as the result of the change: I was enabled to get the manifold away from the cylinder and make a new jointing without lifting the cylinder itself off its seating, and this was something to be thankful for with the thermometer registering several degrees of frost. Having got the parts adrift, I discovered at once the cause of the trouble. The jointing or packing had been made of cardboard instead of a proper heat-resisting material, and had simply been burnt out, leaving practically a free escape not only for the exhaust, but for the in-coming charge as well. My difficulty now was to find something which I might use for packing, and, fortunately, I discovered at the bottom of the tool bag the remains of a tyre gaiter of the kind intended for application inside the cover, between it and the tube, and this served to make a very serviceable jointing. On fixing up everything into place I soon found that the engine had resumed its wonted quietness and power, and I went on my journey blessing the workman who had let me in for this unpleasant experience on a bleak winter's night, and when time, too, was of some importance.

THE LEA-FRANCIS TWO-SPEED GEAR

THE LEA-FRANCIS TWO-SPEED GEAR

The two-speed gear mechanism and clutch fitted to the Lea-Francis motor cycles appears to me to possess considerable merit in its design, being simple and strong with nothing to get out of order. The sectional view on this page shows the general construction. The gear box is carried in the countershaft position on the motor cycle, the drive to the back wheel being by chain. The main shaft carries a double-dog clutch between two toothed wheels, which latter mesh with other toothed wheels upon the lay shaft, and according to which of these clutches is engaged high or low gear is obtained. The main shaft is driven by the engine through a multiple plate clutch, and the double-dog clutch may be moved endwise on this shaft by means of a gear control lever and engaged, as already stated with one or other of the gear wheels mounted freely on this shaft, and which are always in mesh with the corresponding gears solidly mounted on the lay shaft. The chain wheel which drives back to the road wheel is solidly connected to the larger of the two gear wheels on the main shaft, and if the double-dog clutch is locked to the latter, the chain wheel will rotate at the same speed as the main shaft. If, on the other hand, this clutch is locked to the smaller of the two gears on the main shaft, the latter will then drive the lay shaft, which in turn will drive the larger gear on the main shaft at a lower speed than the shaft itself, such a reduction of speed being, of course, a question of the ratios of the gears.

THE REAR LIGHT QUESTION

As the majority of my readers are no doubt aware, the new regulations respecting rear lights on motor cycles have been issued by the Local Government Board, and are to the effect that motor bicycles ridden solo are not required to carry a red rear light, but machines with side cars attached must do so, whilst if the side car is attached to the right-hand or off side, the light must be placed so as to show the full width of the vehicle. Very few side cars are nowadays fixed on the off side, and so in the bulk of cases riders will be at liberty to carry the red light where they please, provided that there is no hindrance to its showing clearly in the direction opposite to that in which the vehicle is travelling.

There does not seem to be much point in the regulation, as other and slower moving vehicles are exempted from its provisions; but I rather incline to the view that all road vehicles should be made to carry rear lights, and I speak after many narrow escapes from collisions with other and slower traffic going in the same direction as myself, owing to the absence of rear lights on the vehicles ahead of me.

BENZOLE AND
EXHAUST VALVE DESTRUCTION

I notice that the correspondence columns of the specialised Press have been swollen of late by letters from motor cyclists who aver that the use of benzole has led to a destructive action on the exhaust valves of their engines, one even going so far as to say that he had to discard a valve after a few hundred miles of running on benzole. I rather suspect the real cause to be general heating up of the engine owing to insufficient air, the use of too large a jet under the altered conditions, or some such reason, and not to the presence of the heavier spirit itself in the charge. I used benzole myself for about nine months in a cycle car engine, and after reducing the jet one size and giving rather more air, I found nothing but advantage from the change. I obtained greater mileage per gallon, the machine climbed hills better, and, of course, there was a saving in the first cost of the fuel itself. The only tendency towards destructive action that I noticed was when filling up the tank. Any splashing of the spirit soon produced a marked effect upon the paint work, and it was not long before a large circle of bare metal showed itself around the orifice through which the tank was replenished. This, however, was a small matter when compared with damage to vital parts of the engine's anatomy, such as valves.

TWO-STROKE ENGINES AND AIR COOLING

It has been stated that, were it not for the low mean effective pressure in a two-stroke engine of the three-port type, it would be impossible to run such an engine air-cooled. This is correct, and it is not to be expected that an engine in which the piston, in addition to performing its own work, does that of the valves also, and in which a ported cylinder is used and the crank-case turned-into a pump chamber, will allow of an M.E.P. equal to that of the four-stroke. For small two-stroke engines with nominal ratings of from 2 to 2 3/4 h.-p., or thereabouts, air-cooling is perfectly successful, provided, of course, that adequate exhaust arrangements are fitted, and this is mainly due, as claimed, to the fact of its being impossible to obtain a higher mean effective pressure. When the makers begin to build two-stroke engines of greater power, it will be seen, unless I am greatly mistaken, that water cooling will become a standard feature, the only alternative being the use of separate pump cylinders, and even then it is by no means certain that the cylinders would not have to be partly, if not wholly, water cooled.

The very essence of light and medium weight two-stroke design is its great simplicity, and if this is going to be removed in any but a very small degree the firm hold which this type of engine is now taking upon the public fancy will be loosened. The majority of newcomers to motor-cycling are willing to give up a good deal in return for simplicity of detail, and this is just where the present design of air-cooled two-stroke engine may claim its highest advantage.

WHY THE ENGINE WOULD NOT START

One of our readers who lives near me, and who owns a three-wheeled cycle-car, came round in a high state of dudgeon one recent Saturday afternoon to say that nothing that he could do would effect starting of the engine. He had not used the car for a week, and when put away after the last run everything was in perfect order. Investigation proved that no spark was taking place; and on examining the contact-breaker we discovered that the rocker arm bearing was a tight fixture, owing to the fibre bushing having swollen. The rocker arm, which carries one of the "points" of the make-and-break, has attached to it, towards the opposite end, a pin which rests in the fibre bearing bush above referred to, and this forms the pivoting point of the arm as it rises and falls to bring the points in and out of contact. All we had to do to put matters right was to ease the pin A down very slightly with emery cloth until it moved easily in the bush B. Some might have thought it preferable to scrape the bush itself on the inside, but as it is a somewhat delicate part, and would easily crack, we thought it better to do the rubbing on the part we knew would stand it, and in any case it was an inexpressible part of a fraction which was removed.

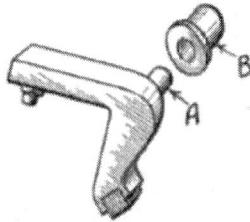

THE TWO-SPEED GEAR MECHANISM
OF THE "CLYNO" TWO-STROKE MOTOR CYCLE

The photographic illustration on this page shows the arrangement of the two-speed gear and clutch as fitted to the "Clyno" two-stroke light-weight. The gears are enclosed in the crankcase, and as one complete unit with the engine. The belt pulley (with which is incorporated a multiple-plate clutch) is mounted on the countershaft behind the engine, as seen; and this countershaft carries at the opposite end to the pulley two gear wheels of which the outer one is larger than its companion.

These gears mesh with others on the engine shaft, the outer pair (one on the engine and the other on the countershaft) provide the low gear, and the inner ones the high gear. The large, outer gear wheel on countershaft runs loosely on its shaft when the high gear is in action, but when it is desired to engage the low gear the inner countershaft gear wheel is caused to slide along the shaft, out of engagement with its companion gear on the engine, and pins formed upon it enter holes in the large gear wheel, thus connecting the two together and locking the large gear wheel solidly to the countershaft, so that the drive is then taken through the outer instead of the inner toothed wheels, and low gear is obtained.

The drive, as will be seen, is thus through either one or the other of the gear wheel sets. If through the outer ones, the engine is running in the low gear; and if through the inner ones, in high gear. There is no chain drive between the engine and the countershaft, as is usually done. A gate change lever alongside the tank actuates rod and lever mechanism, actuating a forked member working in a groove for sliding the high gear pinion to and fro.

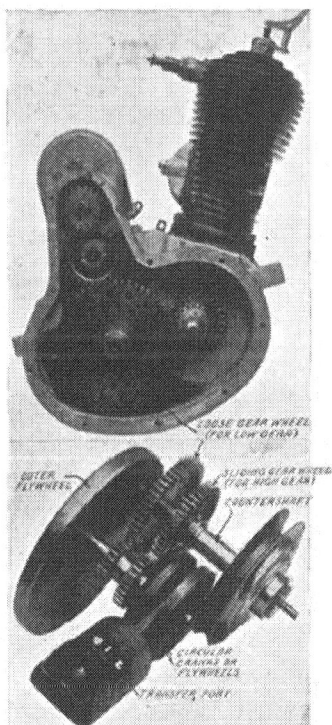

TWO VIEWS OF TWO-SPEED GEAR MECHANISM
OF THE "CLYNO" TWO-STROKE MOTOR CYCLE

A LOOSE TANK CAUSES AN ACCIDENT

A motor cyclist of my acquaintance recently met with a rather nasty accident, owing to the petrol tank of his machine not being tightly fixed in place. He was climbing a stiff hill on his low gear and was leaning forward for the purpose of retarding the spark lever, and must in some way have brought pressure to bear upon the left-hand top side of the tank. The machine is fitted with a three-speed hub type gear, operated as usual by a quadrant lever fixed to the top tube and tank of the machine. The fact of the tank tilting to the left caused the gear rod to move and throw the machine out of gear—that is to put the engine in free position. Being anxious not to stop the engine, the rider left it roaring round, and pushed the tank down on the right-hand side, removing both hands from the handlebars for the purpose. The result was to throw the engine back into gear, and the machine gave such a lurch that the rider was thrown from the saddle whilst the machine fell on top of him with the engine still running. He has since had the tank fixed properly in the frame and advises other riders to see to it that their machines do not suffer from the defect of loose tank fixings.

SOME SILENCER EXPERIMENTS
AND THEIR OBJECTS

Having decided to adopt some means of warming the side car attached to my machine, I looked about for the best way of accomplishing my purpose, and happening across a notice in one of the motor papers of the Thomson-Bennett device I decided to investigate its merits. Accordingly I obtained a set of apparatus specially designed and made by Messrs. Thomson-Bennett, Ltd., of Birmingham, for use with side cars and which comprises a fitting for application to the exhaust pipe, a length of flexible tubing and a foot-warmer adaptable for fixing to the floor of the side car, and devised to act as a silencer, a suitable outlet being provided and made so that a length of piping can be fitted if desired to lead the exhaust gases away aft.

The fitting attached to the exhaust pipe is provided with a butterfly valve with handle, so that the heat may be turned off or on, that is, made to pass through the flexible tubing to the foot-warmer in the side car or allowed to pass straight on to the silencer as usual. I thought some difficulty might arise from the fact that in my engine there are two exhaust pipes to the single cylinder, but to guard against the possibility of not being able to deflect sufficient of the gases to warm the side car, I reduced the area of outlet from the silencer, with the idea that the gases would embrace the opportunity of rushing out through the valve and flexible tubing to the heater, and that taken at a point near the exhaust port and with a practically free means of escape satisfactory results would follow.

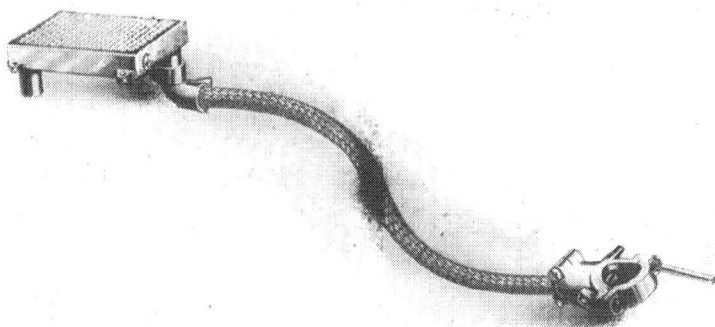

A DEVICE FOR HEATING SIDE CARS

However, when all was fitted up I discovered that it had become impossible to get the engine to two-stroke with regularity, whilst the heater only became moderately warm. I therefore commenced to enlarge the outlet area of the silencer proper by degrees, getting better results each time, so far as regular working was concerned, but with rather a tendency

25

to loss of heat to the side car. I decided, therefore, to try an entirely different method, and having restored the silencer to its original condition, I fitted a butterfly valve in the exhaust pipe to which the heating apparatus is coupled, but lower down, leaving the other exhaust pipe free to discharge into the silencer exactly as before.

This method proved instantly successful. When the heater valve is opened the other one is closed, and vice versa, according to whether I require the heat "on" or "off." As now fitted the side car passenger derives the benefit of the use of a foot-warmer, and the engine runs with greater freedom than before. I found by experiment that a piece of tubing about 9 ins. long, fitted to the footwarmer, greatly improves the working.

The apparatus is very well and accurately made, and can be supplied to fit any make of motor bicycle. My side car passengers speak in high praise of it.

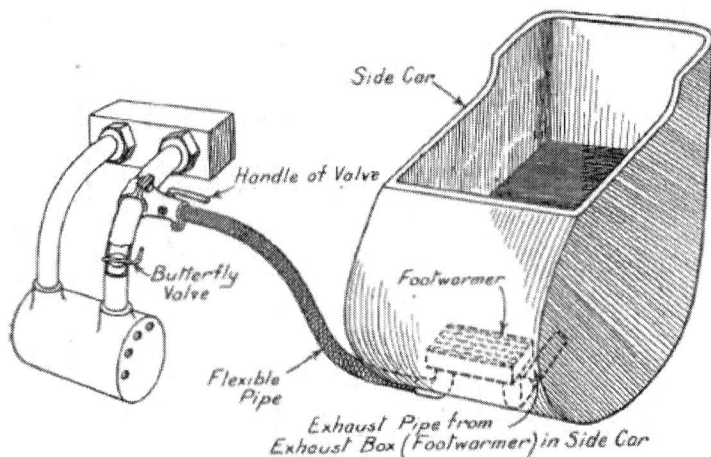

THE SIDE CAR HEATER IN POSITION

A NOTE TO CORRESPONDENTS

I have a number of letters awaiting attention which have been received from readers in many parts of the country and abroad. Although in the aggregate the work entailed in dealing with these must demand a considerable amount of my time being given up, in only two amongst many times that number has even a stamped addressed envelope been enclosed. I always like to assist, so far as I am able, any reader to whom I can be of service, but having as much work of the ordinary kinds on my hands as I can conveniently manage, I am compelled to make a rule, as from now on, that all letters requiring a reply through the post must contain six penny stamps and a stamped addressed envelope. Even then it must often mean that I am putting aside more pressing and lucrative work in order to attend to these enquiries.

TROUBLE AFTER RE-BUSHING

A correspondent asks my opinion as to the poor performance of his motor cycle, which, for the past eighteen months or more, has given him every satisfaction. He says: "The engine has just been rebushed throughout, and at the same time I had the cylinder re-bored and a new piston and rings fitted. Since this work was carried out the engine is hard to. start, and it fails to pick up on the level, whilst its hill-climbing powers are not one-third what they were before the overhaul. It also gets very hot after running a few miles."

It is probably the case, I should think, that either the piston is too tight a fit in the cylinder or that one or more of the bearings are tight. The piston may be a good fit in one part and an incorrect one in another owing to error in re-boring the cylinder. In one case which came under my notice, an engine, after undergoing an overhaul of this kind, refused to start at all although there was a splendid spark and the carburettor "carburetted never so wisely." The trouble was discovered owing to the difficulty experienced in pedalling the engine round. It was due to faulty mounting of the flywheels on the shaft, the flywheels or one of them touching the bottom of the crankcase at one point.

It may be that in my correspondent's case the magneto has been wrongly timed. This would account for the difficulty in starting, sluggish running and overheating, assuming of course that the timing was too far retarded. I have advised him to investigate on these lines, and I feel sure that if he does so, he will discover the cause of the trouble.

LOW-TENSION MAGNETOS

Since the note on the subject of low-tension magnetos appeared in these columns, I have been engaged in a most interesting correspondence with Mr. J. M. Dimond, of Sussex Place, S.W., who wrote tome in the first instance as follows:— "In your 'Motor Cycle Notes' appearing in THE MODEL ENGINEER for January 13th, I notice that you say, in reference to the disappearance of low-tension magnetos on motor cycles: 'It is true that in a few cases low-tension magnetos were fitted to motor cycles of Continental manufacture, but they fell into disuse principally owing to the fact that with this type it is necessary to carry a coil, . . . this being absolutely essential to the production of a sufficiently intense electric spark.'

"From the text," continues my correspondent, "these remarks might be taken to refer to the plain break-spark ignition commonly called low-tension magneto ignition; but, as there are two types of sparks, the low- and the high-tension, of which only the high-tension can be used with an ordinary sparking plug, they must refer to low-tension magneto systems with high-tension coils; the only one of which, so far as I know, to be used on motor bicycles being the Eisemann.

"These systems are not truly low-tension, as the magneto simply replaces the accumulator or battery used in the various high-tension coil systems. You also state that one advantage of the high-tension magneto is that 'Only one high-tension insulated wire or cable is required.' This can hardly be cited as an advantage," remarks Mr. Dimond, "over the low-tension magneto, as the latter only required one wire, and that a low-tension one, which is less likely to leak. The weak point in the low-tension system was of course the tappet-igniting gear, which was apt to get out of adjustment and leakage of compression result. This was partly the cause of its passing out of favour for motor bicycles. However, this point was eliminated by the introduction of such magnetic plugs as the Simplex, Bosch and others.

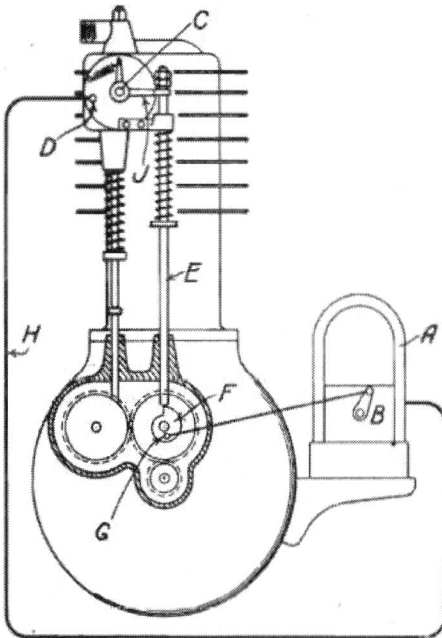

SHOWING LOW-TENSION MAGNETO
IGNITION WITHOUT COIL

"The proof of the reliability of the low-tension magneto is furnished by its continued application to such famous marine engines as the 'Gardner.' The low-tension system," concludes my correspondent "is in no way more complicated than the high-tension, although at first sight it may appear so, owing to its make-and-break being separate from the magneto, and the system you quote as being out of date owing to complication is now appearing on a 1914 motor cycle with self-starting apparatus which is surely evidence of its being up-to-date."

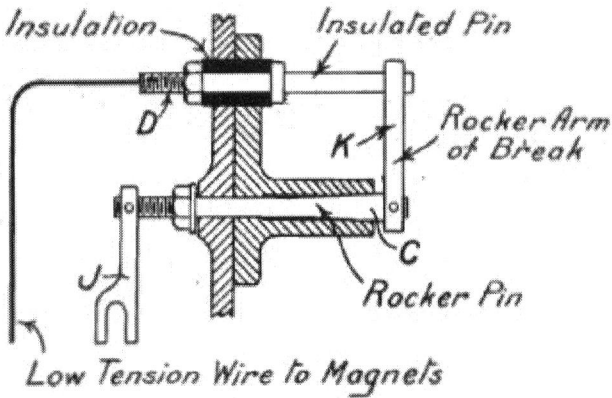

ENLARGED SKETCH OF IGNITER

WHAT *IS* THE LOW-TENSION SYSTEM?

I replied to the foregoing very interesting letter by maintaining that the low-tension magneto system was, as I said in my notes, discarded because it entailed a more complicated mechanism than the high-tension magneto system, in which latter the current is generated and delivered by one self-contained instrument of comparatively small dimensions with a single cable direct to the sparking plug. The low-tension magneto with coil comprises as a system two separate instruments located at a distance apart from one another (as a matter of convenience in carrying) and with two separate wires, one from the magneto to the coil (low-tension) and the other from the coil to the plug (high-tension), and further the adjustment of the separate make-and-break was a matter requiring greater attention than where the contact-breaker is in one with the magneto. I remember very well that some of the N.S.U. motor cycles of date, say, 1905-7, were fitted with the system and that it went against them in effecting a sale second-hand after high-tension magnetos had got properly on the market.

However, it appears that Mr. Dimond in writing to me had an earlier system in mind, namely, the one in which no coil was employed, the low-tension current generated by the magneto going straight to an igniter in the combustion chamber of the motor. This type is forgotten by most motor cyclists, if they or the majority of them ever knew of it, in fact; but it was the true low-tension magneto system after all, and moreover the one which my correspondent was really referring to when criticising my former note. He has been good enough to send me a sketch, which is reproduced herewith, of the arrangement, and he reminds me that it was used by several Continental makers and, in England, chiefly by the Singer. Clyde, and Simms motor cycles and tricycles.

The system works as follows:—A is the magneto, the current being produced by rocking a shield between the armature and the field-magnets by means of the rocker B driven by an eccentric on the cam F. An insulated plug D is so fastened that it projects into the combustion chamber of the motor, and to this plug is fixed a low-tension wire from the magneto. The other pole of the magneto is earthed.

A pin C projects through a stuffing box into the combustion chamber, and is fitted with an arm K which normally rests on D and so completes the circuit. On the outside C is fitted with another arm J. Now when the moment of ignition arrives, the cam F lets the tappet E fall; a projection on the tappet rod strikes J and so lifts K off the pin D, thus breaking the circuit and so causing a spark between K and D. The circuit is then again completed by the cam lifting the tappet and so allowing a spring to pull J back to, its former position. This is the true low-tension magneto ignition. In the ordinary high-tension coil ignition we have an accumulator supplying a low-tension current to a coil which converts in to a high-tension current. Put a low-tension magneto in place of the accumulator and you have the Eisemann system that was applied to the N.S.U. machines. Therefore, concludes Mr. Dimond, I contend that magneto and coil systems are not low-tension ones.

FAULTY BRAKE APPLIANCES

The average motor bicycle suffers, I am afraid, from what a medical rider once termed "Brakitis"—that is, more or less chronic disability of the brakes to perform their allotted functions. I cannot remember ever having owned a machine on which the brakes were in good order for long at a time, and one gets so tired of constantly adjusting and dabbling with the brake mechanism that in time neglect arises, and the consequences of this may be very serious ones. It is a common complaint for rear brakes to stick and refuse to be released unless a vigorous kick is given to the pedal, whilst, as a rule, it is necessary to adjust them frequently to keep them in order; and the means provided do not, in many cases, permit of such fine adjustment as is sometimes necessary. Front wheel brakes applied to the rim are often a great nuisance; the blocks either wear out quickly or else refuse to grip the wheel no matter how forcibly applied, whilst in their case also strict and constant attention is required to maintain them in working order.

I have no doubt that many readers can truthfully, say they never have any trouble with their brakes, which always act properly, but I have never myself experienced any long immunity from brake trouble, nor do I know any other hard all-weather rider who has.

ADJUSTING THE ARMSTRONG GEAR

A Bristol reader, who owns a motor bicycle fitted with the Armstrong 3-speed gear, writes saying that on changing from top to middle gear the engine races a great deal before taking up the drive. "I have," he says, "endeavoured to adjust the gear and have also taken it to a garage, but it is now again slipping and worse than before. It is especially noticeable when changing on hills, not so much in traffic, although even then at times it will slip."

I have advised this correspondent to test the adjustment in the manner advocated by the makers, and which I have already explained in these notes. The lever on the quadrant secured to the frame, and tank of the machine must be pushed forward until a screw or nail can be inserted in the hole provided for the purpose in the quadrant, then, with the handle in this position, the milled nut at the end of the operating rod must be rotated until the back (driving) wheel of the bicycle can be revolved easily by the hand in free position, the belt rim remaining stationary. With this adjustment properly made, and provided there is nothing amiss with the gear or clutch, there is no likelihood of slip taking place. Sometimes belt slip is mistaken for slip in the gear, and my correspondent should make sure on this point.

MOTOR CYCLES THAT ARE "TOO FAST"

I sometimes receive complaints from motor cycling readers that the machines they ride are too fast for them—that is to say, they cannot be run at a moderate speed on the level. These complaints come, as a rule, from elderly riders, or those who are only just beginning to use a motor cycle. One correspondent in a recent letter told me that he had reduced the gear as far as the adjustable pulley would permit, and as he mentions what machine it is, I happen to know that he must be going about driving on a gear of 6 1/2 to 1, and that with a good strong 3 1/2 h.-p. engine. There is absolutely no need to do anything of this kind in order to keep down speed on the level. If the carburettor is correctly adjusted, a proper size of jet used, with choke tube to suit, and the ignition is rightly timed it should be, and, as a matter of fact, is possible to throttle a powerful engine down to a very low speed on the level—of course, with proper handling of the control levers.

To drive on a gear of 6 1/2 to 1 with a powerful engine is not only very undesirable from the point of view of economy, but may lead to actual harm, and especially so if the lubrication is not carefully attended to. The heat generated in the cylinder is much greater than when a suitable gear ratio is employed, and the wear of the belt likewise. From what the particular correspondent I referred to said in his letter, I should imagine that the ignition is somewhat far advanced in his engine, and also that possibly he is using a big jet and high petrol level. If he overhauls these items, and follows some further suggestions I have made in respect of his driving methods, he should meet with better results. The engine in his possession is, as I know, a perfectly docile one in experienced hands.

A SMASH AND THE SEQUEL

A Bournemouth correspondent writes to tell me that whilst riding his 3 1/2 h.-p. motor cycle recently the connecting-rod broke and the piston was forced through the top of the cylinder, the engine being a complete wreck. The machine being practically a new one, he considered that the makers ought to meet him very generously in the matter of cost of the repairs.

They, however, demanded the sum of £8 10s., and accepted no liability of any kind. My correspondent then consulted me on the point, and sent the parts for my inspection. There was an undoubted flaw in the connecting-rod, and it is certain that it was this that caused the accident to take place. At the same time, the makers could not be blamed directly for this, and it was only to be expected that they would endeavour to shelter themselves behind the usual plea that "all precautions that are usual and reasonable have been taken," etc. However, "all's well that ends well," and with a view to settling the matter on the basis of compromise, and to save further trouble, the repairs have been carried out at a cost of £3 5s., or a reduction ot £5 5s. on the original estimate.

FRAMES THAT ARE OUT OF LINE

Motor cycle frames, if tested by means of a straight edge, may sometimes be found to be out of alignment, especially where the machine has been used for hauling a heavy side car. Occasionally, new frames leave the factories slightly out of line, but it is very rare indeed that a frame that is not perfectly true gets past those who are responsible for inspecting or "viewing" the parts without the fault being detected. Such a case, however, was recently brought to my notice, and it was first discovered as a consequence of the belt not running true and steadily in the pulleys. On gauging up the frame it was discovered that the error in the alignment of the frame was a most noticeable one, and the maker's attention being drawn to it, a new frame was fitted without charge.

A NOVEL CONNECTING-ROD

I was recently making a tour of inspection in the Ariel Works, Birmingham, when I noticed, on one of the benches in the engine testing department, a connecting-rod specially adapted for ensuring thorough lubrication of the gudgeon and crank-pin bearings.

As seen in the photograph the big end of the rod has cut in it a number of circumferential grooves, whilst at the small end there is a slot with oil hole. Oil channels are likewise cut in the rod at front and back, communicating with holes in the upper surface of the big end. These combined measures have been taken to ensure vertical lifting of the oil, and its adequate distribution to the two bearings named. The piston on its underside has a cone-shaped projection pointing downwards, and the oil drips from this inverted cone into the groove formed in the small end of the rod, and through the hole to the gudgeon-pin bearing. Similarly the oil on splashing up on to the rod as the milled big end churns round in it, flows downwards and gets to the crank-pin bearing through the hole provided for the purpose. Reference to the illustration will make these points clear. The design is proving highly satisfactory; it is only used on the single-cylinder Ariel machine, not on the twin-cylinder ones.

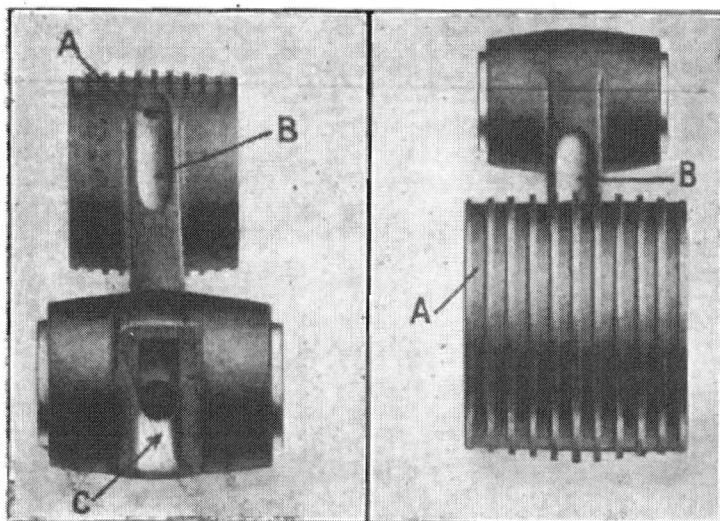

LUBRICATION ARRANGEMENTS ON
THE "ARIEL" CONNECTING ROD BEARINGS

PROTECTING IRON PATTERNS

A COATING for protecting iron patterns from rust is composed of 1 pint of turpentine, 1 gallon of paraffin oil and sufficient of vegetable black to render it thick like ordinary paint. The mass is rubbed on the patterns and, according to the *Brass World*, they have been used over fifty times, when treated with this compound, and still were as smooth as new.

THE ULTRA-LIGHTWEIGHT MOTOR CYCLE

We appear at the present time to be on the threshold of what is likely to prove an immense boom in the demand for the ultra-lightweight or "motor-cyclette" type of machine. I have recently been making enquiries among manufacturers, and visiting the works of several, and everywhere there is a disposition to regard this class of mount as a promising one. Undoubtedly, a very large number of those who would not otherwise take to motor cycling will do so if they can obtain a reliable machine which is easy to handle and control, and which can be picked up, as it were, under one arm and lifted into the house. The difficulty with many—the only difficulty in fact—is the storage question, and as with the motorcyclette this is entirely overcome, the obstacle in the way of their becoming motor cyclists will be removed from the path of the bulk of those who have waited in the hope that the right sort of machine would ultimately appear.

I have examined the specifications of certain machines coming within the category referred to, and many of them are really excellent, the construction being of ample strength and the whole machine extraordinarily light. The price is much below that of the heavier and more powerful models, and altogether the machines are of a very desirable kind. Of course, in this as in everything else, there are grades of excellence, and it would be untrue if I said I agreed with all I had seen in the course of my investigations. In one or two cases the finish left something to be desired, as did also the design of certain minor but important parts. With these put right, however, the machines would be worthy of ranking with the others, and I have no doubt that by the time the present season has advanced a little further the discriminating buyer will have an ample selection of motor-cyclettes from which he can safely make his choice.

MOTOR CYCLE REPAIRS

I am accustomed to receive letters at intervals from readers who have suffered the misfortune of breaking some vital part af their machines and in many cases, owing to the latter being of old pattern, difficulty arises in connection with obtaining replacements. Quite recently I heard from a Bournemouth reader, whose machine is of a make no longer marketed, and it has become impossible to purchase spare parts of the engine, which is some few years out of date. The particular trouble in this instance was the breaking of the connecting rod, and my correspondent appeared to view the position with some concern, as he had been advised by someone whom he consulted that he would have to get a new rod made. I have been able, however, to tender him better advice than this, and as a matter of fact he has since had the old rod repaired at a very slight cost compared with that of having one specially made.

Another reader wanted to know how best to proceed with a fractured cylinder, and he too had been told that it would mean a substantial outlay in cash before the damage could be made good. In this last case the cost would naturally be higher than in the other, but when regarded in comparison with the cost of a new cylinder, it turned out to be not much more than half as much, or a saving of nearly 50 per cent.

These two breakages were repaired in a perfectly satisfactory manner by the oxy-acetylene welding process, and not only at a relatively low cost, but with no loss of time. I was present at a demonstration of this process recently, and witnessed some really remarkable feats in repairing and cutting metal objects. As one instance, a pair of. girder forks belonging to a motor cycle were soundly welded in the space of a few minutes, whilst a water jacketed cylinder, from a motor car engine, was similarly dealt with in a little over ten minutes. The cost in the latter case amounted to just about one fifth of that needed to purchase a new cylinder.

Some tests were afterwards carried out to illustrate the strength and durability of the welds, and it was shown that on submitting the objects, after treatment by the oxy-acetylene process, to bending, and other critical tests, that it was always at the point of welding that the material showed the greatest resistance—in fact, in no instance did the welded portion give out, although the metal collapsed at other points under the tests.

Motor cyclists who require repairs to expensive parts of their machines would do well to make enquiries as to what is possible by means of the process referred to above, before deciding to purchase new parts or have them specially made.

TWO-STROKES AND CARBURATION

It is well known that two-stroke engines are extremely sensitive to slight changes in the carburettor settings, but I doubt if the point is appreciated to the full by the majority of users. I certainly was not aware myself of the full extent of this disposition, although I have been driving a two-stroke and studying its idiosyncrasies for a space of some nine months or more, until during a recent ride I had reason to investigate the cause of indifferent running.

I was due to start on a somewhat long road journey in the afternoon of a recent Saturday, and as the engine had not received any attention for some little time I decided before starting to remove the cylinder and clear away the carbonised deposits on the piston and in the cylinder head and ports. This is but a small task with a two-stroke, but at the same time it is very important that the ports should be clear if really good results are to be secured.

The engine proved remarkably sluggish on the outward run, so much so, indeed, that I was compelled to resort to the middle gear on hills that ordinarily were taken on top. The speed on the level was not materially affected, although the picking up of the engine after even a slight reduction was noticeably poor. I tried a larger jet after making sure that the existing one was not choked, trimmed the sparking plug and contact-breaker points, and made sure that I was getting a good supply of mixture, and an excellent spark to ignite it with. All, however, to no avail, and I was, indeed, hard put to it, when nearing my destination, some half an hour later, to climb the hill up into the town even on the lowest gear.

I determined to get to the bottom of the trouble before attempting the homeward journey. I, therefore, overhauled the carburettor and ignition appliances once again, but found everything apparently in order; also tested the compression both above and below the piston, and found that likewise good. It then occurred to me to try what a slight alteration in the angle of the carburettor float chamber would do, this being easily effected by tilting the whole carburettor on the induction pipe and screwing up the clip.

This proved immediately successful, and I have since been running with the float chamber tilted ever so slightly upward, so that the petrol takes if anything a downward flow to the jet chamber, which is apparently just what was required. Evidently this position had been disturbed during the cleaning operations in the morning, and it was not until it had been resumed that the engine worked at its best again.

It seems somewhat hard to believe that such a slight change of position could make such a pronounced difference in the power given off by the engine, and were it not for the fact that I made certain all was right with the actual working of the carburettor and the ignition apparatus, and carried out a test between each inspection, I should have been driven to the conclusion that I must during the investigations have corrected some fault without knowing it.

TROUBLE WITH A BROKEN VALVE

A Worcester correspondent writes to say that he has experienced trouble on two occasions recently with breakage of the inlet valve stems on his 6 h.-p. engine. The stems break through the cotter opening, and he is at a loss to understand why this should occur. My correspondent says that having no spare valve he was put to considerable inconvenience, and eventually had to travel a distance of over seven miles on one cylinder and in the low gear. Even then he had to run alongside the machine, to which a side car is attached, on every hill.

When valves break through the cotter opening in the steam it usually signifies faulty design of the cam, causing too sharp a lifting motion for the valve. In one case with which I am familiar, breakages of this kind were so frequent that the makers of the engine were obliged to adopt special measures for getting in touch with users of the particular type in which the trouble occurred, and anyone applying for same was supplied free of charge with another design of cam.

Faults of this description do not, as a rule, show themselves on the test bench, where the engine runs under a constant load, and, for the most part, at the same speed. On the road, however, the conditions vary greatly, and the mechanism is subjected to treatment such as does not arise under other circumstances.

STRANGE BEHAVIOUR OF A BELT

Another correspondent, writing from Bedford, complains that the belt on his machine, a new one, turns over on its side on the rim after running a short distance, the effect being to wear the rubber away from the, sides of the belt and cause uneven running. He says he was told at a garage, where he called without the machine, that the trouble arose through faulty alignment of the pulleys, causing the belt to assume a "cross drive" instead of a straight one, and setting up a twisting motion which results in the belt turning over as described.

It is, of course, quite possible that faulty alignment of the pulleys may have something to do with it, but in a new machine there should be little probability of this. Another cause is a warped condition of the belt itself, and my correspondent can test this by first mounting the belt on the pulleys with the name and other marks on the belt reading outwards, and noting how it beds in the back wheel rim; then reversing it so that the marks read inwards, and again noting the fit. It is more than probable that the mere act of reversing the belt will effect a cure of the trouble. That was the result obtained in a case with which I am acquainted. The sketches show the difference in the "fit" before and after reversal in that particular case. When the belt is run down a little it will be found to run well in either direction.

FITTING A NEW PISTON RING

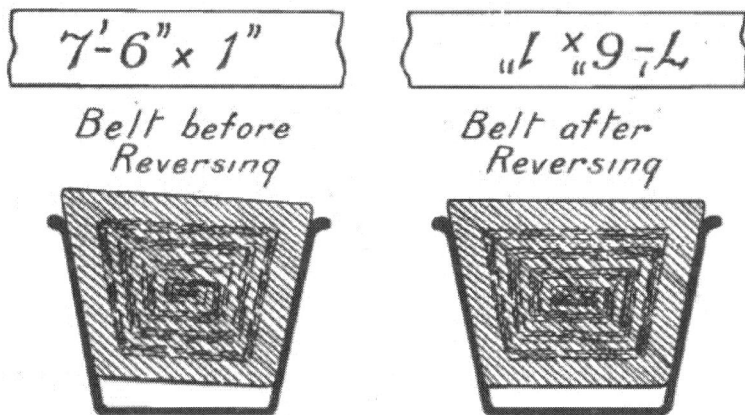

REVERSING A BELT TO IMPROVE THE FIT

The task of fitting a new piston ring is not—at at all events, should not—be a difficult one, yet it sometimes proves the undoing of those who are inexperienced in such matters. I was recently asked to advise a reader of *Junior Mechanics*, who had broken three rings in endeavouring to get one new one in place. As it was in my own district, I called round to see the machine, and found it to be a lightweight with two-stroke engine. The piston is fitted with two rings at the top, and the owner, unable to remove the upper ring, had endeavoured to "spring" the new ring over the top one, the result being as described. After removing the top ring by the knife-blade method—*i.e.*, running the blade of a pocket knife round carefully between the ring and the piston—the new ring was easily mounted in position, and afterwards the top one was replaced. It was then found, however, that difficulty occurred in getting the cylinder over the new ring, and eventually the latter had to be eased at the slot after carefully cleaning out the groove in which it rested.

By these combined methods the new ring was got into place and the cylinder mounted without further trouble; but had the owner of the machine persisted in trying to force new rings over the top one, it is probable that he would have broken many more in his endeavours to do what almost amounted to the impossible. Moreover, there was a decided risk of his breaking away part of the lower portion of the cylinder—a much more serious matter than breakage of a ring.

REAR LIGHTING
FROM THE MAGNETO

Since the order for compulsory rear lighting came into force, motor cyclists have been much troubled in finding a really effective means of carrying the order into effect. I have myself tried oil lamps and acetylene ones—the latter, both with separate generators and supplied from the main or headlight generator—

but have not met with the desired amount of success.

Now, however, I am using an electric lighting set which takes its current from the magneto, and I have found it satisfactory in every way. A brilliant rear light is obtained while the magneto is working, and a separate battery is supplied for use while standing. The system is known as the "Krisco," and it has never given me a moment's trouble. Neither does it appear to affect the magneto in the slightest degree. Altogether I am delighted with the appliance, which is retailed at a moderate figure.

AN OFFICER'S BELT EXPERIENCE

I was returning from Coventry one evening recently, and when about thirty miles from London I fell in with a military officer, the belt fastener on whose machine had broken. The machine—a secondhand one—had been purchased the day previously, and the new owner appeared to be quite ignorant as to how to proceed in order to extricate himself from the situation in which he was placed.

On enquiring whether he had a spare fastener with him, I was informed that he had no idea, and a search in the tool bag revealed the fact that none was there. Fortunately, although not riding a similar machine, I found among the oddments I usually carry—many being relics of past motor cycles—a new fastener of the right size. We, or rather I—next discovered that the half fastener remaining on the belt had had the pin riveted over, the thread presumably having worn. I filed down the riveted portion, and fitted the new fastener in place, the whole business occupying but a short period of time.

The officer admitted that he would not have had the slightest idea of how to proceed, if left to himself, and he was good enough to style me an "expert" for carrying out an operation which the veriest novice is usually capable of performing in the space of a few minutes. Someone to whom I mentioned the matter later was unkind enough to suggest that it was simply a ruse, the purpose of which was to get another to do the work whilst the owner looked on!

A TWO-STROKE QUERY

A Bradford reader writes asking my opinion in the following circumstances. He rides a small two-stroke motor cycle fitted with drip feed lubrication, the feed being adjustable but not visible. There was, he says, a slight leakage of oil round the joint between the lubricator and the tank, and to stop this he screwed the former up more tightly into the tank. Unfortunately, in doing this, the screwed end of the lubricator broke off short, leaving the major part of it in the tank, and rendering the lubricator useless. Some difficulty exists, it seems, in obtaining a new fitting, and my correspondent, having almost daily use for the machine, has resorted to "petrol" lubrication, namely the petrol and oil mixture system. He believes he has got the proportions of oil and petrol correct, inasmuch as there is slight emission of blue smoke from the silencer when running at low speeds; but a decided falling off in power has followed the change, and no variations in the mixture fed to the engine seems to restore the normal conditions. Examination of the carburettor reveals nothing wrong, and the spark is a most excellent one.

On hearing of the circumstances, I wrote asking what steps he had taken with regard to the oil pipe, now disconnected from the lubricator at the top end, but still fixed to unions irt the crankcase and cylinder base, and on receiving his reply to the effect that it was left as it was I advised him to plug the open end at the top in order to prevent air being sucked down into the crankcase, and thus weakening the mixture. This he has since done, and now reports that the engine is working as well as ever.

SIDE CAR ALIGNMENT

I am often consulted on the important point suggested by the heading of this note, and those who write sometimes express astonishment at the poor steering qualities of their machines in view of the "fact that the side car is correctly aligned." Where this is really the case it is practically impossible for the steering to be other than satisfactory, and as a rule it is found, when proper tests are applied, the alignment is not so correct as is supposed.

Quite recently, I was asked to examine a side car combination which was difficult to steer, especially when negotiating right-hand corners. The side car wheel and the back wheel of the machine were found to be entirely vertical, and the tyres on those wheels, in spite of having done over 3,000 miles, showed a perfectly even wear on the treads.

The front-wheel tyre had worn considerably, however, on the near side, the tread pattern having wholly disappeared, whilst on the off-side half of the tyre it was still quite well defined. This showed at once that whilst the side car and bicycle frame were correctly in line, the steering wheel was at fault, and, on testing the spring forks, I found that they had gone over appreciably, and that the wheel, as a consequence, was running out of truth. The forks are capable of being straightened, but they would have first to be removed from the machine. This is now being done, and if the tyre is changed round, so that the near side of the cover comes to the off side, it will tend to equalise the wear of the tread somewhat.

AN EXHAUST LIFTER DIFFICULTY

A correspondent, writing from Dorchester, says that whilst riding his motor cycle late at night he had the misfortune to break the wire which operates the exhaust lifter. "Not having any means of effecting a repair," he says "I was condemned to push the machine, a heavy one, a distance of nearly three miles, of which the concluding half-mile was on an up-grade. I tried to start by pushing over compression, but this was too much for me, and I therefore removed the belt and pushed the machine home. It is. single-geared and no clutch."

There are two methods of "getting going" when the exhaust lifter wire breaks. One is to insert a coin or disc between the tappet and exhaust valve stem, and, after pushing the machine along, contrive in some way to knock the coin out; another is to utilise a strap, cord, or some other means for holding up the lifter whilst a running start is made. I once witnessed a very clever start made without either. The rider mounted his machine at the top of a short decline in the road, and, by furious paddling with one foot whilst holding the lifter lever up by means of an adjustable spanner, he managed to get the engine started. If by any method the exhaust valve can be held off its seating until sufficient way can be got on the machine, and the valve then dropped, a start can be made, and it is quite feasible to get going by starting the engine up "on the stand"; then, after mounting the saddle, let the wheel drop to the ground whilst revolving rapidly, opening the throttle as the wheel touches the ground, when machine will dart forward, and can, with luck and a little skill, be kept going without the engine being stopped. This is done frequently by testers in Coventry and elsewhere.

INTERRUPTED IGNITION

A correspondent, writing from Halifax, tells me that he was recently held up on the road with ignition trouble of a kind which he was unable to diagnose, the result being that he was condemned to push his machine, a 2 1/2 h.-p. medium weight, a distance of 1 1/2 miles. The machine was running well he says, until, when just at the foot of a hill, the engine ceased firing.

My correspondent says that when this occurred he "tried the tank for petrol and examined the carburettor to see that it was working," and continues: "I then became convinced that it must be something to do with the ignition, but on removing the plug 1 found that I had got an excellent spark. I accordingly replaced the plug, and, to my satisfaction, the engine started up at once on the stand. When, however, I endeavoured to get going again on the road it would not fire except for a few spluttering explosions just at first. I again tried the spark, and tested the cable terminal as well as the sparking plug, and each time there appeared a really good spark.

"A second attempt to proceed met with the same result as the first; and as it was dark and inclined to rain, and, furthermore, as I was only about one and a half miles from home, I decided to waste no more time, but to push the machine home. I have not had a chance yet of looking into the matter; but I shall be going home again soon, and would like, if possible, to have your opinion with regard to the matter."

Having been thus consulted, I can only suggest that the trouble arose from a fault connected with the insulation of the high-tension cable. If in some way—as, for instance, through resting against the hot cylinder—the insulating rubber became destroyed, thus baring the wire, and this particular portion of the cable rested against the cylinder or other metal object, the result would be shorting of the current, and, of course, the engine would cease firing. When the cable was removed from its normal position for testing, the bare wire would be taken out of contact with the metal, and sparking would be resumed.

Thus, when my correspondent tested the sparking plug and cable with the machine on the stand he would get a good spark, and in replacing the parts he evidently, and quite unconsciously, caused the cable to fall clear, but on lowering the wheel to the ground it must have returned to its old or another similar position with the wire in contact with metal. The "few spluttering explosions" referred to occurred because the contact was only partly completed, but the vibration probably shook it back into the original place, and the "shorting" again became permanent.

As a rule when this happens there is a smell of burning rubber to guide one in tracing the cause of the trouble; if, however, the insulation becomes cut or worn through, this, of course, does not occur. I was once held up on the road in a similar way, but on that occasion I found the trouble to be due to a "stuck up" carbon brush, the spring of which had got jambed.

COUNTERSHAFT GEAR ADVANTAGES

A letter recently to hand bears upon the point as to what constituted the advantages of the countershaft type of gear box for motor cycles. The writer of the letter says that he often sees it claimed that such a form of gear is superior to others, and also that it is instrumental in prolonging the life of the belt where such is used. He is, however, at a loss to know wherein the superiority lies, or why the countershaft principle should favourably affect the belt.

The countershaft position for the gear lends itself favourably to motor cycle construction for the reason that it is convenient, the gearbox being located in the best of possible positions—namely, between the driving and the driven mediums. The gearbox may be of unobtrusive proportions, and yet provide three different "speeds" or ratios of gearing. The parts are, or should be, few in number and the construction throughout of ample strength. The dog clutch principle of gear changing can be conveniently employed, and there is no difficulty in incorporating a slipping clutch with the mechanism, having either hand or foot control. The position is also suitable for a kick starter without the need of a separate casing.

Owing to the fact that the pulley is mounted on a countershaft, and the latter driven by chain from the engine shaft, the speed at which the countershaft runs in relation to the revolutions of the engine can be increased; or in other words, the pulley shaft can be geared down, and for this reason a larger diameter of flanges can be used. Hence the favourable effect upon the belt which does not have to pass round such a small arc as would be the case if the pulley were mounted directly upon the engine shaft. The weight of the gear box is carried in a more favourable position, and for these combined reasons the countershaft gear position is nowadays regarded as the most suitable one for the purpose.

DANGER WITH SIDE CARS AT CORNERS

A well-known manufacturer of motor cycles, whose output includes a powerful side car combination, has expressed the view that the cause of the wheel of a sidecar lifting when being driven round a left-hand corner at too high a speed is. centrifugal force, and that this force comes from the motor cycle itself and very little from the side car and its passenger. A 5 h.-p. or 6 h.-p. motor cycle, such as is used for passenger work, usually weighs anything from 250 to 280 lb., and even more in some cases, and added to this is the weight of the rider. When this weight, while travelling at a high speed, is suddenly turned to the left, centrifugal force throws the weight of the machine and rider to the right and it is this force which lifts the side car wheel from the ground.

When a solo rider takes a corner at high speed he naturally leans the machine over in the direction in which he is turning, and thus throws the centrifugal force in a downward direction at an angle to the ground. The greater the speed the greater must be the angle; and the only danger to a solo rider in these circumstances is when the angle of the machine to the road is so acute that the tyres cannot retain their grip, and a side slip results.

This cannot occur in the case of a side car for the reason that the latter holds the machine in a vertical position, and nothing can be done by the rider to counteract the effect of centrifugal force, other than by throwing the weight of his body as far over in the direction of turning as is possible.

A READER'S LAMP ENQUIRY

A Hampstead reader writes, asking my opinion in the following circumstances; "Acting on the information previously given by a motor cycling authority, he says, "I soldered up the movable connections of my motor cycle headlight to comply with the new regulations which prohibit the use of lamps capable of being swivelled—or, to use the official language, moved independently of their fixings on the machine. I gathered from what you said that this was sufficient to meet the requirements of the authorities; but whilst riding my machine through a small country village one recent evening I was stopped by a special constable, who, after examining the headlight, stated that it did not comply with the law, and that I was, therefore committing an offence. I told him that I had seen it stated in print that soldering was sufficient, but he persisted in his view, and took careful note of my name and address, the number of my machine, time, date, and place, and I can only suppose that he means to make trouble for me. Will you please say whether you think I can be prosecuted or not, and, if so, what defence I can put up."

I do not think my correspondent need trouble his mind about the matter at all. The special constable was probably the village baker or some such person, with no more knowledge of what is or is not the law's requirement on the point than has the village pump, and if he applies for a summons he will probably be told by those who know better than he himself that no offence has been committed. The largest lamp makers in England have been in touch with Scotland Yard on the subject, and have elicited from the authorities there the fact that if the movable parts are soldered, so as to render them fixtures, the lamp may be said to comply with the regulations. This should be quite enough for any "special," or, for the matter of that, ordinary constable in the land; and in the unlikely event of a prosecution arising out of the case of my correspondent, I shall be glad to supply him with a copy of the letter in which the lamp firm inform me of the Scotland Yard dictum on the point.

BELT WEAR ON EXPANDING PULLEY GEARS

Another correspondent, whose letter is dated from Preston, Lancs, asks what it is necessary to do in order to obtain the large mileages from the belt used on a motor cycle fitted with the infinitely variable type of gear such as we often read about in the motor cycle Press. He says that he is unable to get more than about 2,000, or, at the most 2,500 miles, from a single 1-in. belt on a machine having a 6-h.-p. engine and hauling a side car; whereas it is not infrequently stated in the correspondence columns of the papers that mileages of from 5,000 upwards are being secured even with 8-h.-p. engines and heavy side cars.

He asks whether it is necessary to remove the belt at the end of each run, and what other steps there are which will conduce to better results than those he himself is obtaining.

I have had in use for nearly a year an 8-h.-p. side car combination fitted with belt transmission. I secured 5,320 miles from the first belt, and the second has run over 3,000 miles and shows but little sign of wear. I do not remove the belt from the pulley at the end of each run, but I am very careful to "gear up" before putting the machine away; that is to say, I leave the pulley in its most favourable position for the belt, namely, on top gear, which in my case means that the belt is bent round an 8-in. diameter circle and at an easy radius; whereas did I leave the pulley expanded it would mean that the belt was bent round a very small circle, and this would tend in time to crack the edges and otherwise harm the construction.

Something depends, of course, upon the rider's handling of the gear control mechanism, and sudden changes of gear should be avoided as much as possible. The movements should be as gradual as circumstances permit, and full advantage at all times be taken of the infinitely variable principle on which the gear operates. Many treat the gear as though it had only two or three definite changes, and this is not the way to get the best wear out of the belt. On my own particular machine, which has a 1 1/8-in. belt, 9 ft. in length, running over an 8-in. pulley placed on the countershaft, conditions are primarily favourable to longevity of the belt, but even so, unless care is taken in the directions indicated, it is quite possible, if not altogether easy, to wear a belt out in the course of a much smaller mileage than those quoted. My correspondent does not say what belt he uses, but presumably it is one of the well-known makes.

A QUERY RELATING
TO TWO-STROKE WORKING

The Editor has handed me a letter containing a query *re* two-stroke working, and relating principally to the lubrication of the engine. The writer says that he recently acquired a two-stroke lightweight, and according to the maker's instructions there should be a light film of oil vapour issuing from the tail pipe of the exhaust-box at low speeds. This, however, is not apparent on his machine, although it never seems to lack oil and the engine revolves with perfect freedom, and, indeed, the whole machine is thoroughly satisfactory. He would be glad of an explanation of what he terms a phenomenon, there being no signs of under lubrication, unless the non-appearance of the oil vapour may be taken as such.

In a machine of this character, which relies upon invisible drip feed for the lubrication of the engine, the only visual sign that all is as it should be in this respect is the appearance of the slight blue haze at the silencer openings or extremity of the tail pipe; but even though this does *not* appear it does not necessarily indicate that the engine is being starved of oil. Still, if the pipe is clear, and no obstruction exists at any point in the lubricating ducts, it is difficult to see how the non-appearance of the vapour is to be accounted for, except on the supposition that the amount being fed is rather below that which—though, perhaps, not normally demanded by the engine—represents a safe margin to work upon. If the rider whilst driving the machine along suddenly raises the exhaust lifter, and as smartly releases it again, he will probably be rewarded with a sharp popping sound and the emission of a cloud of blue smoke, and this will show him that at least no danger is being incurred. I think that were the machine mine I should investigate with the purpose of finding out why it is the symptom referred to by the makers is not apparent, and try the effect of materially increasing the rate of oil feed, also noting the mileage in oil consumption that the engine runs on.

MAKING AN ARMATURE FOR A MAGNETO

I recently spent a most interesting couple of hours in a magneto factory in the Midlands, and there witnessed the production of an armature from start to finish. The core, or body portion, of the armature is made of cast-iron shaped thus X, and the ends are first of all rough milled, after which soft iron laminations and malleable iron ends are fixed in place. Screwed rivets are employed for securing these in position. The cores are next taken to the lathe and are there turned in preparation for the final milling operation on the core and sides. After this the core is drilled and tapped, and then it is grooved to allow of clearing the primary winding. The winding operation is performed in a separate department, and partly by means of special machines, the primary winding being done by hand and the insulating material fixed in place before the machine starts on the secondary winding. After this the armatures are finally bound and "taped," and then they go in a stove for the purpose of getting all moisture out of the tape. The latter is afterwards covered with several coats of shellac until the armature presents a firm and glossy appearance.

The armature after being removed from the stove is allowed to cool and is then taken to the test shop, where a reading is secured of the ohms resistance. Next, the armature is returned to the machine shop for final turning ready for the fitting of the inner ball race, and after another visit to the "view room" for further examination it passes to the assembling department, where it is tested under pressure before being placed in position in the magnetos. During the test referred to the magnetos are run at varying speeds, ranging from 200 to 3,000 r.p.m., under a pressure of 120 lb. per square inch.

A HINT TO USERS OF SMALL TYRES

A recent experience of mine has taught me a lesson from which I shall certainly endeavour to profit in the future. I was unfortunate enough, whilst riding my 2 1/4-h.-p. Levis two-stroke, to run over a large piece of broken glass, the result being that the back tyre, already somewhat worn, received a bad gash, which I found on examination to be incapable of repair by ordinary methods. As it happens, I live in a country district some 50 miles from London, and it is a matter of the greatest difficulty to procure motor cycle replacements locally, and I knew from the start that it would be impossible to purchase such an article as a 24-by-2-in. tyre, this size not being easily procurable anywhere out of London, and not always there. I therefore telegraphed to a firm in town who stock practically every class of motor tyre, and asked them to send forward a 24-in. by 2-in. cover and tube by passenger train immediately, hoping to, in any case, receive the goods in the course of the same day. They, however, had not come to hand five days later, and I was put to some inconvenience by being deprived of the use of the machine, which forms a very convenient means of transit to and from the main line station, three miles away. In the end I called upon the vendors in London, and was there shown a despatch note proving conclusively that the tyre had been sent off as soon as it was possible to procure one of that size—that is to say, two days after the receipt of the telegraphic request for same; and for the rest, it had been reposing in the parcels office at the London terminus, although marked "Very Urgent," and consigned by passenger and not by goods train.

The moral to be pointed from the incident is that all users of 24-by-2-in. tyres—and with the growing popularity of the two-stroke lightweight the number is an ever-increasing one—should take the precaution of providing themselves with a spare cover at least, and, better still, a complete tyre; for should the necessity ever arise, as in my own case it has done, of replacing the tyre at a moment's notice, much delay and disappointment will probably arise before the new tyre can be obtained.

THE CAUSE OF A PETROL STOPPAGE

A reader of these Notes writes me from Welling-borough respecting an experience he has recently undergone with his machine, and which, he says, might have landed him in an annoying predicament but for the kindly intervention of another motor cyclist who happened to pass by. The machine had been running well, says my correspondent, on the previous trip; but when he took it out for a run to a neighbouring town in the evening he found that the carburettor was difficult to flood, and that when started the engine lacked vigour and could hardly be made to climb a hill that it at other times simply plays with. "I am riding my first motor cycle," says my correspondent, "and thought I was more or less prepared for any eventuality, and when I found the engine was performing so badly I put it down to a choked jet, and accordingly stopped and examined the part referred to, finding a small particle of foreign matter lodged at the orifice of the jet. Removal of this did not, however, improve matters, and I began to think that the ignition was at fault, when a motor cyclist drew alongside and asked if he could be of any assistance to me. I detailed the circumstances to him, and he advised an examination of the gauze at the base of the float chamber. I was not aware prior to this that such a gauze existed, but on its being exposed to view we found quite a large piece of fibrous material lodged between the gauze and the mouth of the petrol pipe, this effectually stopping the bulk of the petrol from getting through. We were both at a loss to understand how matter of this description could have worked its way into such a position."

The machine is one in which the oil is mixed with the petrol in the tank, and I can only conclude that, provided the petrol itself was poured through a strainer, the foreign matter got carried into the tank along with the lubricating oil, and from there found its way into the petrol pipe and eventually on to the gauze. It often happens that matter of this and similar descriptions lie on the tank itself or the can from which the oil is being poured, and unless great care is taken it is quite an easy matter for it to get swept into the tank, and then trouble is sooner or later almost certain.

It is always advisable when in doubt as to the petrol supply to "feel" the carburettor by agitating the plunger of the float chamber. If the chamber appears to be empty, it may be taken as a certain indication that something or other is preventing the petrol from flowing from the tank to the carburettor, it being assumed, of course, that the rider knows there *is* petrol in the tank.

GETTING THE LAST DROP OF PETROL

Apropos of the subject of petrol flow referred to in the preceding Note, there can be little doubt but that a lot of hard work is sometimes performed by motor cyclists in pushing their machines, when, as a matter of fact, the thing can be avoided. It seldom happens that the last drain of petrol leaves the tank in ordinary circumstances; there is usually a small quantity swishing about at the bottom which cannot in normal conditions reach the carburettor. In many cases the quantity is a not inconsiderable one—say, even about a pint—in a wide tank having the bottom only just covered with slight "wash" of spirit. If the filler cap be removed, and the machine run up a bank to cause the petrol to flow toward the rear of the tank, and the rider places his mouth to the orifice and blows into the tank, it will be found that the float chamber can be filled, the operation being repeated if necessary after the machine has again come to a standstill.

By this means I have on more than one occasion succeeded in covering distances of from one to two miles, and once managed to reach a town three miles distant owing to the last half-mile or so being downhill. Whilst blowing into the tank the finger should be kept on the float chamber, and directly the petrol is felt to be flowing in the engine should be started up and the machine put in motion. I have known motor cyclists push their machines some distance, and then on the matter being put to the test the engine has been set going by the method referred to. One filling of the float chamber is generally sufficient for a run of about half a mile, or, perhaps, a little more, if the road be favourable.

A QUERY RESPECTING
MOTOR CYCLE LICENSING

A reader of *The Model Engineer*, residing at Shoreham, is in a quandary respecting the law as to the registration of motor cycles, and writes as follows:—"Owing to circumstances which I could not foresee," he says, "I find myself in the position of having a motor cycle, and not being able to find the money to register it or pay the licence to keep it. The machine has been 'left' to me by a relative who has gone to the front, and I am told that I must not ride it until I have had new numbers assigned to it and paid what my informant calls the 'carriage tax.' Will you please tell me how I stand in the matter?"

If the machine is already registered, as it may be assumed it is, and has numbers assigned to and carried by it, then all the new owner need do is to apply to the authorities to have the numbers transferred to him, and for this he will have to pay one shilling. It is usual for the County Council to require advice of the application for change from *both* the old and the new owner; but, presumably, if the circumstances were explained to them they would be content to transfer the registration without this formality, so far as it concerns the late owner. As regards the licence—that is to say, the Inland Revenue tax of £1—I should advise my correspondent to wait until the first day of next month (October), when it will be possible to obtain the licence for the remainder of the present year for half the amount chargeable if the licence be granted prior to October 1st. The new owner must, of course, procure his driving licence (5s.), and also have the numbers transferred before he essays to ride the machine on a public highway.

INACCESSIBLE MOTOR CYCLE PARTS

Readers of these notes often write me complaining of the inaccessibility of certain parts of their machines, and particularly the engines. Some motor cycles do certainly suffer from this defect, and it arises in part, if not entirely, from the designer's endeavour to provide a compact construction, with as little waste of space as possible. The thing is often overdone, however, and much inconvenience caused thereby; and it behoves everyone, when selecting a machine, to bear in mind carefully the point as to accessibility, and it is wise to examine the machine thoroughly before purchasing in this particular respect. Chain adjustments, engine head room, and magneto position are three points which require attention on the part of the prospective purchaser; and in the event of any detail being obviously difficult to get at, enquiry should be made of the vendor as to whether any special tool or appliance is provided in the kit to render the task easier of accomplishment.

It is hardly fair to criticise, as some do, side car machines, parts of which are difficult of access from the near side owing to the presence of the side car. It is obviously impossible to avoid this, and the best must be made of matters in such a case. One of our readers complains that when he wants to examine the contact breaker of his magneto, and especially when desiring to alter the ignition setting, he has to "hang over the saddle, head downwards, between the cycle and the side car to see when the points break." This is quite unnecessary, as by holding a small pocket mirror opposite the contact breaker he can, by turning the armature shaft in the ordinary manner, see by reflection exactly when the points break, and, without assuming anything in the character of an uncomfortable position, the arm can be slipped through the frame, and the mirror held in the left hand, whilst with the right one the armature shaft is moved until the correct position is attained.

A RECENT HEADLIGHT DECISION

How very necessary it is to observe with the utmost strictness the new regulations respecting movable headlights on motor cycles is shown by a case that was recently reported in the daily Press.

The motor cyclist was summoned (in Bedfordshire) for "having upon his motor cycle a lamp which was capable of being moved (swivelled) in an upward or downward direction, independently of its fixings," or some such formula.

In fining him £1 and costs, the magistrates informed the defendant that it was "within their power to confiscate the complete machine," and enquiry shows that this *is*, as a matter of fact, the case. In this particular instance only the lamp itself was confiscated, but even then it represented a considerable loss to the owner, inasmuch as it was of an expensive pattern and only recently acquired. The machine itself was. a high-class side car combination, so that had it been taken the matter would have assumed a very serious aspect indeed for the defendant.

I hope none of my readers will fall victims to circumstances in this way, and if they do it will really be their own fault after the warnings they have had. Fancy the remorse one would experience at losing ones entire machine, all for the sake of neglecting to carry out a trumpery operation involving a half-penny worth of solder.

DAMAGE TO A PISTON

A Kettering reader informs me that when recently he took the cylinder of his motor-cycle engine off he discovered a small pear-shaped hole in the lower part of the piston. "I am quite certain," he says. "that there was no sign of this when last I had the cylinder down, and I am rather anxious about it. Is it likely to become enlarged, and if so, what will be the probable result?"

It is difficult to account for a thing of this kind if it is known certainly that the piston was perfectly sound before. Sometimes a blow hole in the casting develops into an opening of this kind after a period of use, but not unless it is practically through the metal from the start. In the case of such a discovery as that made by my correspondent, the part should be carefully cleaned and examined for signs of cracking. If none are apparent, it is hardly likely, that the damage will spread; but if even a slight crack is discovered there will be a possibility, and even a probability, that it will do so.

As the opening is in the skirt of the piston and below the rings, its presence will occasion no harm so far as the working of the engine is concerned; but the owner should keep a watchful eye upon it, and examine it from time to time in order that any development of the defect may be noted as quickly as possible. It is sometimes a good plan to drill a clean hole slightly larger than the one already, formed. This will prevent the one that is absorbed from spreading.

HOT INDUCTION PIPE

It will sometimes happen that after a period of use, and particularly when the machine has been worked somewhat hard and continuously, that the induction pipe of the engine gets very hot; and in a twin engine, where the pipe is short and bridges the space between the cylinders, this is especially noticeable.

When this occurs it is a sure sign that the cylinder or cylinders require cleaning for carbon deposits, and it will be found on examination that the inlet valves require attention, the ports being partly blocked by burnt matter, and the valves themselves in need of grinding. The induction pipe should never get unduly warm, and if it does it is certain that the above is the cause.

A NEW MOTOR CYCLE RECORD

A motor cyclist in America has performed the feat of riding a distance of 1,594 miles in three days, nine hours, fifteen minutes. He had only seven hours' rest throughout the time, and consumed his "meals" in the saddle—often, as he says, at speeds of from 40 to 50 m.p.h. The machine was a 7-9 h.-p twin cylinder one, as designed for the 1916 season

A MOTOR CYCLE "LIGHTING" CASE AND ITS SEQUEL

My readers will, I hope, excuse me for taking up a large amount of the space allotted to me in this week's issue in dealing with what, to a large extent, is a personal matter. Inasmuch as it has a wider and more generally applied significance, however, I thought it would be as well to make the information known for the guidance of, if not as a warning to, others who number themselves among readers of the *M.E.*

The plain fact of the matter is that I have fallen a victim to the new laws which have been framed in connection with the lighting of motor cycles under the Defence of the Realm Act, and just at a time when I have been particularly busy in pointing out to others, both in these Notes and elsewhere, how careful they should be in conforming to the rules.

The circumstances in which I tumbled across my fate were such that I can only consider it as a case of predestination— that is to say, one in which the saying "what is to be, is to be" applies.—and there is no getting away from it. I was driving an 8 h.-p. Zenith side car combination through the town of Bedford on the evening of Sunday, October 3rd, when I was stopped by the police, the outcome of the interview being that I was informed I should be reported for contravention of the lighting laws, it being alleged that the rays from the head lamp projected upwards instead of horizontally, thus apparently, in the opinion of the police, being likely to assist hostile aircraft.

As a matter of fact, the lamp had been giving no small amount of trouble for some time prior to reaching Bedford, and on arriving at the outskirts of the town I closed the water regulating valve altogether and left it in that position up to the time of being stopped. The trouble had been water in the tube and burner, and the difficulty was to maintain anything like an adequate light in the circumstances, let alone anything in the nature of an excessive one. Nothing was said by the police at the time about the light being too brilliant, the complaint being, as I have already said, that it projected upwards instead of straight ahead.

THE LAMP AND ITS FIXINGS

The lamp was quite a moderately-sized one, and when, some months back, the new laws came into force, I fitted a smaller burner than that supplied by the makers, so that in all the circumstances I felt absolutely safe, especially as it is obligatory to proceed at a very low rate of speed through the town, owing to the habitually crowded state of the thoroughfares on Sunday evenings. The lamp was fixed in a perfectly horizontal position, and the movable parts had been carefully soldered in order to make them rigid, and consequently to relieve me of any responsibility in connection with the non-swivelling order. It came, therefore, something in the nature of a shock when I was peremptorily called upon to stop, my licence demanded, all particulars taken, and the information about being reported conveyed to me. I pointed out to the sergeant of police that the light was a very low one, and also that the lamp was occupying a horizontal position on the handlebars; but these facts, apparently, had no weight with him at all, and I could clearly see that his intention was to carry the matter further. Both prior to reaching the town, and again after leaving it, I met motor cars carrying headlights of 100 per cent. upwards greater illuminating power, and, indeed, saw one in Bedford itself just before being stopped. It was therefore all the more inexplicable to me why I should have been singled out for particular attention by the police. I mention all these details simply because I want to impress upon my readers how extremely necessary it is for them to use the utmost caution in connection with the use of lamps on motor cycles at the present time.

THE SUMMONS AND THE HEARING

In due course the summons was delivered at my home address, this being about a fortnight subsequent to the date of the "offence." I found that what I was to be charged with was that I, "being a person driving a motor car—to wit, a motor cycle—unlawfully did carry thereon a light of greater brightness than was necessary for the public safety, contrary to the Order as to lights in the borough of Bedford made under Regulation 11 of the Defence of the Realm Regulations, 1914." Here, then, was a new state of affairs, for what I had to answer to was having used a light of "greater brightness," whereas I had expected all along to be summoned for having one that threw its light up too high. However, when, in due course, the case came on for hearing, the police in their evidence laid more emphasis upon the angle of the light than upon its brightness, and insisted that it had constituted a danger in the sense conveyed by the provisions of the Act.

They also alleged that I had taken no steps to modify the light when its illegal nature had been pointed out to me, and, generally speaking, they used every endeavour to secure a conviction. In giving my evidence I emphasised the points referred to in the second paragraph of these Notes, and also pointed out that the same lamp had been used continuously on the same machine for a period of over twelve months, in London, Coventry, and Birmingham, etc., where the lighting restrictions are particularly severe; that it had never once been complained of, and that had I attempted to reduce the light further the effect would have been to practically extinguish it altogether, or, at the least, to render it absolutely useless for the purpose of assisting me in controlling the machine.

I was cross-examined at some length on behalf of the police, and it was pointed out to me during the cross-examination that the power existed to confiscate the entire machine and its equipment.

At the conclusion of the hearing, the Bench, comprising a chairman and six other magistrates, retired, and remained out of Court for a period of upwards of half an hour. After they had left, the Clerk informed me that in such a case as this the magistrates had the power to inflict the following penalties, namely:—

(1) Six months' imprisonment.
(2) A fine of £100.
(3) A combination of the two.

Added to which there was the further possibility of the machine being confiscated. I admit that the period of waiting for the judgment of the Court—that is, the half-hour during which the magistrates were absent—constituted a somewhat trying time in view of the possibilities that had been intimated to me; but when at last the members of the Bench reappeared the sentence was quickly made known, the Chairman, without any introductory remarks, stating: "The defendant is convicted; fined two shillings and sixpence."

If I say that I considered myself very fortunate in getting off so lightly it is not in the sense that I admit I was to blame, for to that point of view I am still as emphatically opposed as ever; but, considering the powers that are vested in those who try these cases, and realising that the evidence on which they have to base their judgments must of necessity be of a conflicting nature, the result of such a case must always be in doubt up to the end.

I certainly do congratulate myself on having obtained a perfectly just and impartial hearing; and if I may offer any advice to those who may unfortunately find themselves in a like position, it is that they (1) appear in person and conduct their own defence; (2) give their evidence in as concise a manner as possible, endeavouring to avoid all side issues; and (3) produce the lamp in Court. By following these rules it is practically certain that the Court will be impressed by the defendant's *bona fides*, and will give him the benefit of all reasonable doubts. That, however, is what I conclude must have happened in my own case.

NEW TWO-STROKE ENGINE DESIGNS

A well-known firm in Birmingham is experimenting at the present time with a two-stroke engine, the twin cylinders of which are set in accordance with the well-known V style so commonly adopted in four-stroke practice. For the purposes of test, two of the standard single cylinder engines have been utilised, these being mounted V fashion in the frame at an angle of about 90 degs., and the suitable connections made. In its final form the engine will have a common crankcase, and the design of the frame will be somewhat modified. Some ingenuity was required to overcome ignition and carburetting problems, but these now seem to have been entirely solved, and later I shall hope to give further details of what, to my way of thinking, is an advance in the right direction where two-stroke practice is concerned.

In another case a two-stroke engine has its cylinders arranged in the horizontally opposed manner and water-cooling is applied. The cylinders each comprise two separate diameters, so that virtually it is a four-cylinder engine, whilst at the same time only one common crank-pin is used. They are both most interesting designs, and well worthy of further consideration in these columns. This it is intended to afford them.

SHIFTING CARBURETTOR LEVERS

"H. S. J.," of Middlesbrough, writes to let me know that he is experiencing trouble with his carburettor control levers, which, he says, automatically close as he rides along. He further states that he has tightened the nut on the top of the "capstan," *i.e.*, the circular part on the handlebar in which the levers move, but to no avail. Examination of the air and throttle slides in the carburettor spraying chamber shows these to be in perfect order, no binding or other hindrance occurring; yet it is impossible to keep the levers open unless the pressure of the fingers is kept against them.

This, my correspondent remarks, "is a great annoyance, and I am at a loss to understand the cause of the trouble or find a remedy for it. "The throttle lever is the worst," he goes on; "it simply jumps back to the closed position if I do not hold it open, and the only position it will stay in for more than a couple of seconds on its own accord is that of full open, and then, unless the air lever is also pushed open wide, it automatically jumps back again."

The action of the levers depends upon friction in the guides to maintain them in any one position that may be desired. The necessary degree of friction is evidently absent in the case referred to, and no amount of turning of the head of the bolt will produce it in circumstances, for it is evident from what my correspondent says that the bolt is tightened up as far as is possible already without producing the desired result.

The remedy is to remove the capstan fitting from the handlebar, and with a light hammer tap all round the under part that is below the lower groove in which the throttle lever works.

If this does not effect an improvement, a small punch may be employed and a series of indents formed in the base, around the path of the lever movement. This will have the effect of closing in the groove somewhat or, at least, of slightly restricting the space, in which the lever works, and thereby set up the small amount of friction required. Some twelve months back I experienced the same trouble, and I overcame it in the manner described.

THE ADVANTAGE
OF SPRING FOOTBOARDS

Only when one has experienced the discomforts that attend the use of rigid footrests or footboards on a motor cycle, in the course of long continuous runs over bad or indifferent road surfaces, is it possible to really appreciate the benefit to be derived from spring or flexible footboards. These are easy to make, and if the machine is already fitted with two pairs of rests—one in front, and the other in the bottom bracket position—the two can be connected by means of some stout flexible material, allowing a sufficient amount of sag to provide the necessary leg reach. I have had fitted to my lightweight machine, a pair of such rests, the material used being balata belting about 8 ins. wide. Aluminium plates, capable of being moved in a forward or backward direction to allow of shifting the position of the feet, are provided, and the whole arrangement is exceedingly comfortable and effective.

I have made some long rides since the arrangement—which, by the way, forms the basis of a patented invention—was fitted, and I can only say that it has made all the difference irr the world to the comfort of riding the machine. There is no chattering of the feet, and all vibration up the legs is entirely done away with. The material on which the fluted aluminium plates rest and slide slopes from the higher level of the front support to the lower level of the back ones, and in such a manner as to form an efficient leg guard, which keeps off the cold and wet in a very satisfactory manner.

A FLEXIBLE FOOTBOARD

I can recommend anyone who suffers from foot and leg vibration to try an arrangement of this sort, for, if my own experience goes for anything, they will find in it a complete remedy for the trouble.

An alternative is to mount an ordinary type of footboard on springs. I tried this, but did not meet with anything like the same good results as from the flexible anti-vibratory boards I am now using. With the latter it is quite possible to stand up on the plates or mount by means of them, thus proving their stable character.

A NEW MOTOR PROPELLED BICYCLE

There might not appear to be much to choose between a motor bicycle and a motor propelled bicycle; but my readers will doubtless be able to appreciate the difference. The latter is a "pedal" cycle with an engine to propel it, and in one of the later designs the engine is located at the rear of the saddle, being supported on a specially constructed carrier, and driving the road wheel by means of a belt.

The accompanying sketch clearly shows the arrangement, and it will be noticed that the engine is placed in a horizontal or lying down position, with the cylinder pointing in a forward direction. A chain from the engine shaft actuates the two-speed gear placed more towards the rear and this chain engages on its underside—that is to say, where it is slack—with the sprocket on the armature shaft of the magnet thus actuating it, and thus producing the necessary current for ignition purposes. The final drive is by belt from a pulley on the gear shaft to a belt rim on the back wheel.

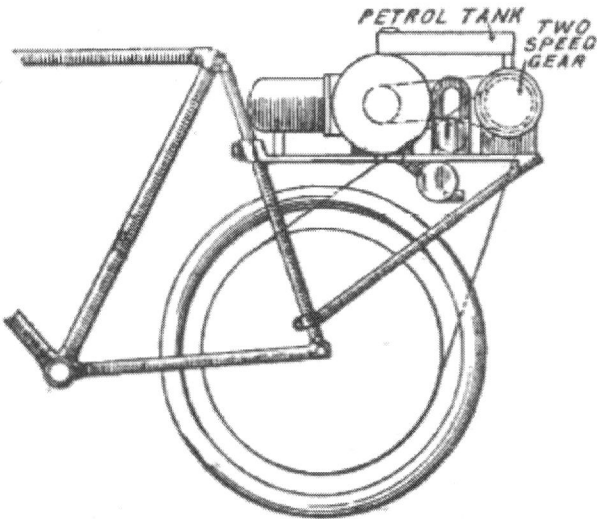

MOTOR MOUNTED AT REAR OF MACHINE

Bicycles constructed on this principle are already being constructed, and it certainly seems to be to me a plan containing some merit. The weight of the engine, etc., being above the back wheel will keep the latter down, and, as every motor cyclist knows, there is less vibration to the rider when the carrier is well loaded than otherwise, although a dead load is one thing perhaps, and a working power unit another. The drive is a direct one, and the only thing I should be doubtful about, would be the method of actuating the magneto. However, it remains to be seen how it works in actual practice. The machine is said to have shown itself capable of reaching a speed of thirty miles per hour under ordinary road conditions.

MY RECENT "LIGHTING" CASE

I would like to take this opportunity of cordially thanking all those who have been good enough to write offering me their congratulations on the result of the above case. Their kindness is very greatly appreciated.

Printed in Great Britain
by Amazon

35041357R00043